BOOKWORMS

Creepy Crawlers

Grasshoppers

Anika Abraham

Cavendish Square
New York

Published in 2019 by Cavendish Square Publishing, LLC
243 5th Avenue, Suite 136, New York, NY 10016

First Edition

Website: cavendishsq.com

Library of Congress Cataloging-in-Publication Data

Names: Abraham, Anika, author.
Title: Grasshoppers / Anika Abraham.
Description: First edition. | New York : Cavendish Square, 2019. | Series: Creepy crawlers | Audience: Grades K-3. | Includes index.
Identifiers: LCCN 2018021937 (print) | LCCN 2018023032 (ebook) | ISBN 9781502641915 (ebook) |
ISBN 9781502641908 (library bound) | ISBN 9781502641885 (paperback) | ISBN 9781502641892 (6 pack)
Subjects: LCSH: Grasshoppers--Life cycles--Juvenile literature.
Classification: LCC QL508.A2 (ebook) | LCC QL508.A2 A25 2019 (print) | DDC 595.7/26--dc23
LC record available at https://lccn.loc.gov/2018021937

Editorial Director: David McNamara
Editor: Kristen Susienka
Copy Editor: Nathan Heidelberger
Associate Art Director: Alan Sliwinski
Designer: Megan Metté
Production Coordinator: Karol Szymczuk
Photo Research: J8 Media

The photographs in this book are used by permission and through the courtesy of: Cover Luciano Queiroz/Shutterstock.com; p. 5 CasarsaGuru/iStock; p. 7 Image by Chris Winsor/Moment/Getty Images; p. 9 Rujira Bumrungkarn/Shutterstock.com; p. 11 Poravute Siriphiroon/Shutterstock.com; p. 13 Gail Shumway/Photographer's Choice/Getty Images; p. 15 Kurt_G/Shutterstock.com; p. 17 Klimek Pavol/Shutterstock.com; p. 19 Vincent Grafhorst/Minden Pictures/Getty Images; p. 21 Nixx Photography/Shutterstock.com.

Printed in the United States of America

Contents

A grasshopper looks creepy.

It has big eyes and a long body.

Grasshoppers start life inside an egg.

A female grasshopper lays eggs in the ground.

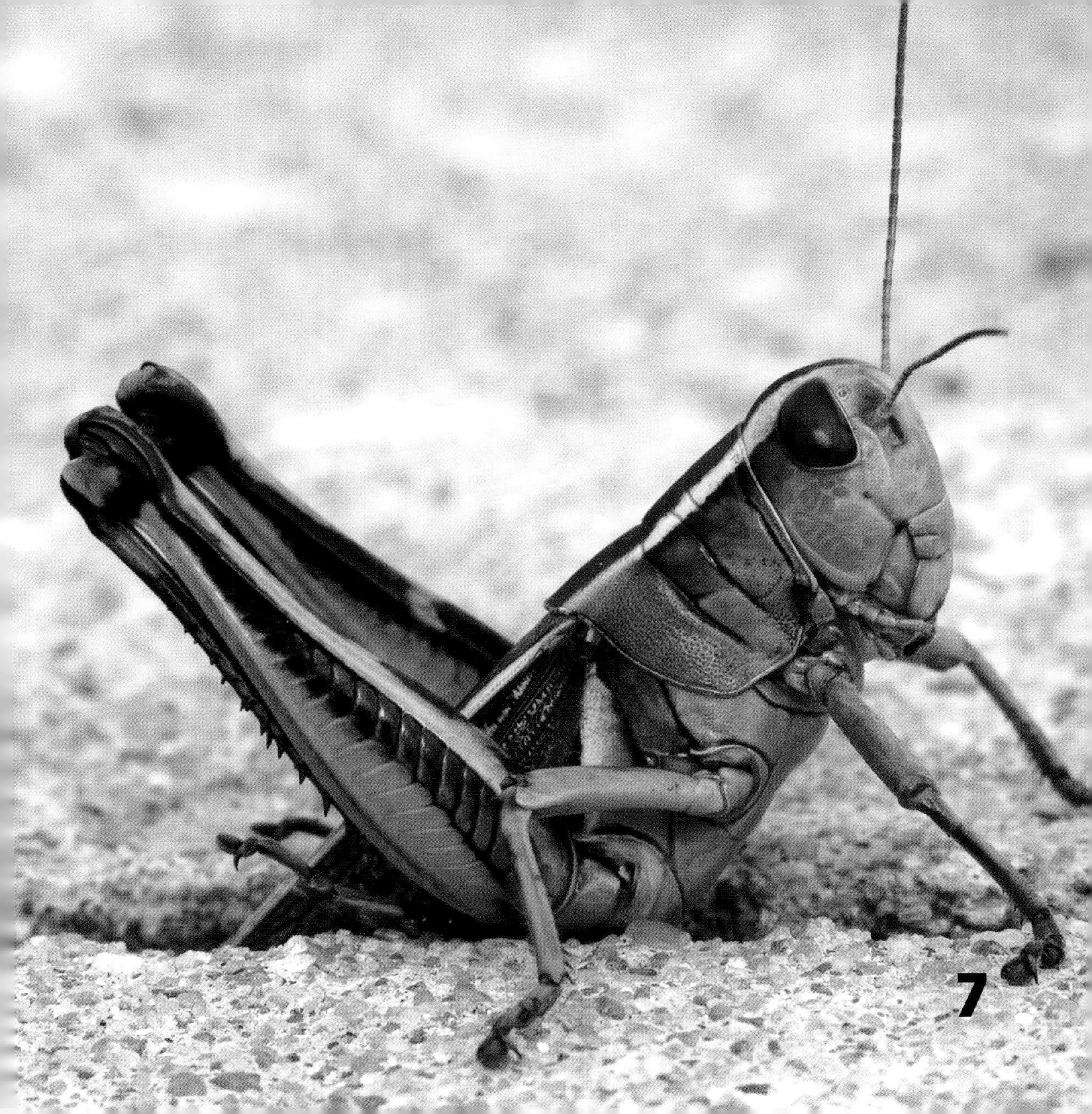

The egg **hatches** into
a grasshopper.

A young grasshopper is
called a **nymph**.

Grasshoppers can be big or small.

They can be brown, green, or gray.

A grasshopper has **antennae** on its head.

Antennae help a grasshopper feel things around it.

Grasshoppers have wings.

Wings help them fly.

Their wings can have bright colors on them.

Bright colors help keep **predators** away.

Grasshoppers have ears that help them hear.

But a grasshopper's ears are on its belly!

Grasshoppers live in leaves or grassy areas.

They like to eat food that grows there.

A grasshopper has strong back legs.

These help it hop through grass.

That's how a grasshopper got its name!

New Words

antennae (ann-TEN-ay) Organs that grow out of an insect's head.

hatches (HATCH-ez) When a baby animal or insect breaks through its egg.

nymph (NIMF) A young grasshopper without wings or with much smaller wings than an adult.

predators (PRED-uh-torz) Animals or insects that want to eat or hurt other animals or insects.

Index

About the Author

Anika Abraham enjoys the outdoors. She likes learning and writing about animals, insects, and plants. She lives in Chicago, Illinois, with her cat Scrooge and her dog Thunder.

About BOOKWORMS

Bookworms help independent readers gain reading confidence through high-frequency words, simple sentences, and strong picture/text support. Each book explores a concept that helps children relate what they read to the world they live in.